AF228574

ANIMAL ECO INFLUENCERS

PRAIRIE DOGS

Builders on the Plains

RACHAEL L. THOMAS

Consulting Editor, Diane Craig, M.A./Reading Specialist

Super Sandcastle

An Imprint of Abdo Publishing
abdobooks.com

ABDOBOOKS.COM

Published by Abdo Publishing, a division of ABDO, PO Box 398166, Minneapolis, Minnesota 55439.
Copyright © 2020 by Abdo Consulting Group, Inc. International copyrights reserved in all countries.
No part of this book may be reproduced in any form without written permission from the publisher.
Super SandCastle™ is a trademark and logo of Abdo Publishing.

Printed in the United States of America, North Mankato, Minnesota
102019
012020

THIS BOOK CONTAINS
RECYCLED MATERIALS

Design: Kelly Doudna, Mighty Media, Inc.
Production: Mighty Media, Inc.
Editor: Liz Salzmann
Cover Photographs: iStockphoto, Shutterstock Images
Interior Photographs: Alamy Stock Photo, pp. 20, 21; Getty Images/Dorling Kindersley RF, pp. 8, 9;
Getty Images/iStockphoto, pp. 6, 7, 8, 10, 11; Mighty Media, Inc., p. 11; Shutterstock Images, pp. 3, 4, 5,
6, 7, 8, 9, 11, 12, 13, 14, 15, 16, 17, 18, 19, 21, 22, 23

Publisher's Cataloging-in-Publication Data
Names: Thomas, Rachael L., author.
Title: Prairie dogs: builders on the plains / by Rachael L. Thomas
Other title: builders on the plains
Description: Minneapolis, Minnesota : Abdo Publishing, 2020 | Series: Animal eco influencers
Identifiers: ISBN 9781532191886 (lib. bdg.) | ISBN 9781532178610 (ebook)
Subjects: LCSH: Prairie dogs--Juvenile literature. | Prairie dogs--Behavior--Juvenile literature. |
    Grassland animals--Juvenile literature. | Prairie animals--Juvenile literature. | Animal ecology--
    Juvenile literature. | Wildlife habitats--Juvenile literature.
Classification: DDC 599.367--dc23

Super SandCastle™ books are created by a team of professional educators, reading specialists, and
content developers around five essential components—phonemic awareness, phonics, vocabulary, text
comprehension, and fluency—to assist young readers as they develop reading skills and strategies and
increase their general knowledge. All books are written, reviewed, and leveled for guided reading, early
reading intervention, and Accelerated Reader™ programs for use in shared, guided, and independent
reading and writing activities to support a balanced approach to literacy instruction.

CONTENTS

ECO INFLUENCERS

What is an eco influencer? It is an animal that can change its ecosystem. All members of an ecosystem affect one another.

Beavers build structures that create **wetlands**. The wetlands support other wildlife.

Ochre sea stars prey on mussels in **tide pools**. This allows more of other animals to live in the ecosystem.

Prairie dogs are eco **influencers**. They dig tunnels in their prairie **ecosystem**. This creates healthy soil and shelter for other animals. Prairie dogs are builders on the **plains**!

PRAIRIE DOG
TOWNS

Prairie dogs live throughout the Great **Plains**. They live in small families called **coteries**. Many coteries make a ward. And many wards make a town.

A prairie dog coterie

TOWN SIZE

At one time, there was a town in Texas that covered 25,000 square miles (65,000 sq km). About 400 million prairie dogs lived there! But most prairie dog towns are much smaller than that.

A prairie dog town

BURROW BUFFS

Prairie dogs dig **burrows** underground. Some have tunnels that are 100 feet (30 m) long! Prairie dogs spend a lot of time working on their burrows.

Burrows have many **entrances**. Below each entrance is a small space used to listen for danger above ground.

Burrows have rooms for different activities, such as sleeping and using the toilet!

Prairie dogs build mounds at each entrance. They stand on the raised dirt to better scout for danger.

Other animals often use prairie dog burrows. These include black-footed ferrets, snakes, and owls!

Burrowing owls

MOUND
LISTENING ROOM
TUNNEL
SLEEPING ROOM
TOILET
THINK!
What do you do in
the different rooms
of your house?

GREAT PLAINS IRRIGATION

Prairie dog tunnels are waterways across the Great **Plains**. When it rains, water flows underground through the tunnels. This prevents flooding above ground. It also helps **irrigate** the soil.

Rainstorm in
the Great Plains

Sometimes, rain causes the tunnels to flood. But prairie dogs are **expert engineers**. They include side rooms in the **burrow** that are safe from rising water!

THINK!

You use water in your home every day. Can you imagine where it **drains** to?

UNDERGROUND OVERGROUND

Prairie dogs are natural **fertilizers**. Tunneling adds air to the soil. Prairie dog **urine**, **feces**, and fur get added too. These make the soil healthier.

Healthier soil allows healthier grasses and plants to grow. Bison, elk, and other animals eat the healthier grasses in prairie dog territory.

Animals such as bison eat grasses on the Great Plains.

GRASS NIBBLERS

Prairie dogs work like little lawnmowers! Sometimes, they nibble an area clean of plants. Insects then move to the bare areas of ground. Birds eat the insects.

Prairie dogs eat grasses, roots, and seeds.

READY FOR DANGER

Prairie dogs leave their **burrows** to find food. So, their bodies are built to help them stay safe in the open **plains**.

PRAIRIE GUARD DOG

When above ground, at least one prairie dog watches for danger. If a predator appears, the lookout warns the others. Prairie dogs use different calls for different predators.

Excellent hearing to listen for danger before leaving the burrow
Eyes on the sides of the head. This makes it easier to see predators coming from any direction.
Sharp claws and teeth for fighting
Strong hind legs to stand up straight while on guard
THINK!
How do your senses, such as vision, hearing, and smell, keep you safe?
15

FOOD WEB

Prairie dogs also affect their **ecosystems** through the **food web**. They provide food for many predators.

Prairie dogs are the only food source for black-footed ferrets. The ferrets also use prairie dog tunnels for shelter. Without prairie dogs, black-footed ferrets cannot survive in the wild.

Black-footed ferret

PRAIRIE DOG PREDATORS

As prey, prairie dogs support these animals in the **ecosystem**.

NO PRAIRIE DOGS?

Prairie dogs are important eco **influencers**. But human activity has put prairie dogs in danger. Prairie dogs are often seen as pests to be trapped or killed.

Prairie dogs are protected in places such as Theodore Roosevelt National Park.

FARMING

Farming is another danger for prairie dogs. In the 1900s, the prairie dog population fell. One main reason was more prairies being turned into farms. Today, many people are trying to find ways for prairie dogs and farmers to live together peacefully.

UNCERTAIN FUTURE

Prairie dogs can get sick with a type of **plague**. This illness can kill most of the prairie dogs in a prairie dog town. They put the medicine in food. Then they scatter the food where prairie dogs can eat it. Scientists hope that the medicine will save the prairie dogs.

The medicine tastes like peanut butter!

ECO INFLUENCER FACT SHEET

Common name: Prairie dog

Class: Mammal

Life span in the wild: 3 to 8 years

Population trend: Threatened

Diet: Herbivore

Size in relation to humans:

FUN FACTS

Prairie dogs are related to squirrels.

Prairie dogs are not related to dogs. Early settlers named them prairie dogs because their calls sounded like dogs barking.

Male prairie dogs move from **coterie** to coterie. But females stay in the same coterie for life.

PRAIRIE DOG
QUIZ

1. Prairie dogs live in forests.
   True or **false**?

2. Which predator depends most on prairie dogs to survive?
   A. golden eagle
   B. coyote
   C. black-footed ferret

3. What caused the prairie dog population to fall in the 1900s?

Answers: 1. False 2. C 3. Farming

GLOSSARY

burrow—a hole or tunnel in the ground that is used for shelter.

coterie—a family of prairie dogs.

drain—to flow out.

ecosystem—a group of plants and animals that live together in nature and depend on each other to survive.

engineer—someone who knows how to build strong, useful structures.

entrance—a door or a way in.

expert—a person very knowledgeable about a certain subject.

feces—solid bodily waste.

fertilizer—something used to make plants grow better in soil.

food web—the feeding relationships between different organisms in a community.

influence—to cause something to change.

irrigate—to supply land and soil with water.

plague—a disease or event that causes deaths or harm to many.

plain—a large, flat area of treeless land.

tide pool—a shallow area of water left near shore when the tide goes out.

urine—a clear, yellowish liquid that is made by the kidneys and released from the body as waste. It is also called pee.

wetland—a low, wet area of land such as a swamp or a marsh.